TALIA'S CODEBOOK

FOR MATHLETES

MARISSA MOSS

WALKER BOOKS

First edition 2023

Library of Congress Catalog Card Number 2022907029
ISBN 978-1-5362-1802-2

23 24 25 26 27 28 LEO 10 9 8 7 6 5 4 3 2 1

Printed in Heshan, Guangdong, China

This book was typeset in Patrick Hand Regular.
The illustrations were done in ink, watercolor, and gouache.

Walker Books US
a division of
Candlewick Press
99 Dover Street
Somerville, Massachusetts 02144

www.walkerbooksus.com

TO ELLERY, WITH THANKS FOR INSPIRING DASH

This journal belongs to:

TALIA ZARGARI

I'm starting this notebook to help me figure out middle school. Getting things out of my head and onto the page always helps me think—or at least it's fun to do. Problems seem smaller once you can draw them. And 6th grade is full of problems!

MIDDLE SCHOOL IS <u>NOT</u> WHAT I EXPECTED.

Sixth grade was supposed to be great—a new, big school, getting a locker, having real science classes and labs instead of just a handout on caterpillars becoming butter-flies. The end of being a little kid!

ELEMENTARY SCHOOL: HOW IT IS		
MATH	SCIENCE	ENGLISH
$1+1=?$ Baby problems, like 1 + 1 = 2. Or worse, memorizing the times tables—again!	Plant a seed. wait. wait. wait. wait. Finally, a sprout!	Read a book. Sometimes good, sometimes boring.

MIDDLE SCHOOL: HOW IT IS		
MATH	SCIENCE	ENGLISH
$\frac{7}{12} \div \frac{3}{5} \times \frac{8}{3} = ?$ Complicated, challenging problems—finally more math fun for me.	Build a robot. IS MY COM-MAND YOUR WISH	write . . . and publish your own story!

Some of that stuff is true, but there are other things about 6th grade that aren't so great. One thing I didn't expect at all.

THE THINGS YOU DON'T WORRY ABOUT
CAN TURN OUT TO BE THE WORST THINGS OF ALL.

My (maybe) best friend

For the first couple of weeks of middle school, everything was fine with us, just the way it should be. Dash was in most of my classes, so I always had someone to sit with.
   Then this happened:

First I was shocked. Then I got mad.

I knew I shouldn't yell. I mean, who wants to be friends with someone who screams at you? But I couldn't stop myself.

It felt like Dash was ashamed of me, that I embarrassed him in front of his other friends. And the way I was
screaming, I <u>was</u> pretty embarrassing. Luckily we were
walking home, far from other people. I could see Dash was
miserable. But I was even <u>more</u> miserable.

I told him that he was talking the way a grown-up does
when they try to convince their kid something awful is
really neat, like getting a shot or going to the dentist.

Calling something rotten a secret doesn't turn it into something cool. And this was definitely NOT cool.

I started imagining lunch all by myself, not having an instant partner in our shared classes, not walking in the hallways together.

I was deep in imagined misery, not even noticing where I was walking, when Dash grabbed my arm.

He's right about that! And he was making it even trickier. I needed my best friend around more than ever.

It was our first big fight since 2nd grade, when I acciden-
tally broke his new Nerf gun and hid it in his closet because
I was too embarrassed to admit what I'd done.

I was so ashamed, I finally had to tell him.
This seemed a lot bigger.

DASH WAS DESPERATE.

All week, I couldn't stop thinking about what Dash had said . . . and how he had looked.

THE DASH SPECTRUM

HAPPY DASH

NORMAL DASH

SAD DASH

MAD DASH

DESPERATE DASH

OTHER KIDS CAN BE REALLY MEAN!

In elementary school, it wasn't that hard to fit in. If you bothered to brush your hair and teeth (and sometimes even if you didn't), kids would still want to play with you. But now it feels like everyone is judging you, like all kinds of weird things are being measured and rated. It's about more than being popular—it's about being simply good enough, "normal," whatever that means, or "regular," as Dash put it.

GRAPH OF NORMALITY

GREEN	normal, fit right in
BLUE	a little odd, maybe one thing that sticks out, like a haircut, could be good or bad
YELLOW	can't put your finger on it, but definitely not normal
ORANGE	3 strikes and you're out—too odd for most things
RED	total weirdo—nobody will ever go near you!

NORMAL KID

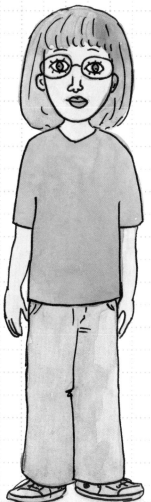

ELEMENTARY SCHOOL

In elementary school, the questions were easy! Yes or no:

Can you jump rope?

Ride a bike?

Tell good jokes?

Laugh at other people's jokes?

Know how to share?

Are you fun to be with?

MIDDLE SCHOOL

In middle school, the questions are tricky, and it's impossible to know the right answers!

Are you wearing the right clothes?

What do you post on EYEgram?

How many likes do you get?

Do you have the right friends?

The right enemies?

Are you cool?

What makes someone cool?

All of this leads to one big deduction.

(A deduction is something specific you figure out by observing a lot of general information.)

DEDUCTION #1

MIDDLE SCHOOL ISN'T ABOUT LEARNING STUFF
FROM TEACHERS AND BOOKS. IT'S REALLY ABOUT
LEARNING HOW TO GET ALONG WITH OTHER PEOPLE.

I'm pretty good at the book-type stuff, not so good at the
people stuff. Reading a person is much harder than reading a textbook or a math problem.

When I took a seat next to this
girl in my homeroom class, she
made this face.

Umm . . .

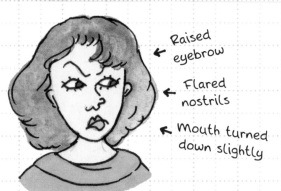

Raised
eyebrow

Flared
nostrils

Mouth turned
down slightly

The chair was empty. I thought it was free.
I wasn't sure how to read her face. Was she disgusted at
how I smelled? Disapproving of how I was dressed? Disappointed I took a seat she was saving for a friend? Whatever
it was, she made me feel like I'd done something wrong.
But I hadn't shoved her or anything. Still, her glare made
me want to disappear!

The funny thing is, I've actually studied expressions for my drawing—particularly eyebrows. They can tell you a lot about how a person is feeling.

All I can say for sure is that nobody ever looked at me that way in elementary school. Even on days when I forgot to brush my hair. I know I'm not a cool kid, but I'm not a freak, either. I'm just ordinary. At least, I thought I was . . .

So I kinda get why Dash is so worried about fitting in. I'm not facing the same pressure (except from the mean-faced girl), since my other friends think it's great that Dash is my best friend (plus they say he's cute). I guess it's different for boys. Other boys think having a friend who happens to be a girl is a mark against you. If I had to deal with that, what would I do? Would I keep a distance from Dash so I wasn't marked as even weirder than I already am?

READING EYEBROWS

If you draw faces with the eyes and mouths all the same and just change the eyebrows, you can see how expressive eyebrows can be. Much more than mouths even because people can control their mouths. But how many people can control their eyebrows?

Eyebrows don't lie!

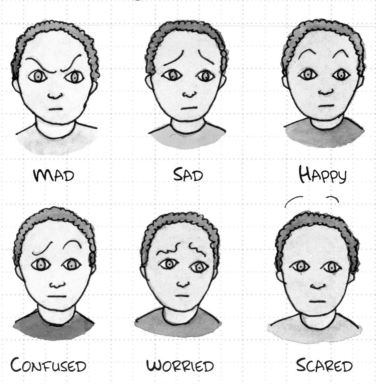

MAD   SAD   HAPPY

CONFUSED   WORRIED   SCARED

IT'S HARD (IMPOSSIBLE?) TO SEE YOURSELF
THE WAY OTHER PEOPLE DO.

Does Dash really want to stop being my friend because I'm a girl or because I'm me? Am I embarrassing? Just to him or to everyone? Am I too nerdy or awkward? Too different?

I keep thinking of the way that girl looked at me when I sat down next to her. It's not a good feeling.

It's like I have a glob of spinach between my front teeth. Or toilet paper stuck to my shoe.

When I look at myself in the mirror, I still have the same curly hair, the same green eyes, the same olive skin I've always had. If I haven't changed, why am I suddenly not friend material?

A GOOD FRIEND HELPS YOU SEE YOURSELF CLEARLY—BECAUSE THEY SEE YOU CLEARLY.

I was too embarrassed (and scared) to ask Dash the REAL reason he didn't want to be my friend anymore. What if it turned out to be way worse than just some stupid peer pressure from a bunch of dumb boys? What if I'm boring? That's one of the worst things a friend could be. Or selfish. Or rude.

I don't think I'm any of those things, but what if I'm wrong? I was too afraid to ask Dash, but I could ask Skye and Luisa, my other friends.

SKYE

we've been friends since 3rd grade.

Like me, she's always writing and drawing stuff she notices.

Sometimes we have draw-offs, contests where we each draw a quick comic and see which is best. It's not a real contest since she always wins. I'm just not a good artist, no matter how much I try.

OUR LAST DRAW-OFF: DANCING MICE

BALLET

MINE

BALLROOM

SKYE'S

(my drawing of Skye's drawing, which is why it looks like I did it, not Skye)

LUISA

We've been friends since 5th grade, when Luisa moved here from Texas. She's funny and smart and super organized. She's the queen of lists—even her lists have lists inside them, like those Russian nesting dolls. Plus she's super generous. You need something, she'll give it to you. I've borrowed socks, books, markers, even pajamas from her.

Skye and Luisa like Dash and Dash likes them, but since he lives right next door, I see him more than I see them (well, I used to). Plus I've known him forever. He was my first best friend and I thought I would never outgrow him. Now it looks like he's outgrown me.

The week after Dash "broke up" with me, I couldn't keep it secret anymore. I finally told Skye and Luisa what happened.

I held my breath for a loooooooooooooooooong time.

Phew! They had me worried for a minute there. I was thinking I shouldn't have asked them anything.

OBSERVATION # 7

SOMETIMES EVEN GOOD FRIENDS CAN MAKE YOU MAD. BUT THEN THEY CAN TURN AROUND AND MAKE YOU VERY HAPPY.

Enough with the cafeteria ladies! I thought you wanted our advice about Dash. Isn't that what you're really asking about?

Of course it is. You think she really cares about what kind of friend she is? She cares about what kind of friend Dash is!

That's not true!

I mean, I do care what kind of friend Dash is, but I'm really wondering if maybe I'm not interesting to him anymore.

OBSERVATION # 8

FRIENDS WILL ALWAYS HAVE ADVICE FOR YOU.

The question is, is it good advice or terrible?
It's not always obvious.

SKYE'S ADVICE

Just give Dash some space.

He's basically a good person, and I bet he ends up missing you and comes crawling back, asking you to forgive him.

In the meanwhile, you still have us, remember?

LUISA'S ADVICE

Yeah, Skye and me should count for something!

You can always find another boy to be your friend and prove to Dash that it's fine for boys and girls to be friends.

But really, why bother when you have Skye and me, the two greatest friends ever!

You can see how this advice may be a bit biased. They DO have a point that they're wonderful friends. But so is (was?) Dash.

I'M TERRIBLE AT THESE KINDS OF DECISIONS!

Luisa and Skye both made a lot of sense, but it's really hard for me to accept that the solution to a problem is to DO NOTHING. I always feel better if I can try something, even if I fail. I hate the idea that there's nothing I can do, that this is all in Dash's control. But I can't MAKE him be my friend. Even I know that's not how friendship works.

FRIENDSHIP RECIPES

MIX AND BAKE
AND YOU'LL HAVE INSTANT FRIENDS!

RECIPE A: How grown-ups think friendships work.

- Add two kids the same age.
- Mix together.
- Get instant friends (as if!!).

Talia, this is Cedric, my boss's son. You two go play in your room while we talk. He's six, just like you. You'll have a lot of fun!

I bet if I told you to go play with Ms. Layfield down the street because you're both thirty-eight, you'd tell me I was being ridiculous!

Cedric was NOT fun! He was mean and bossy and broke my favorite paintbrush. It's not easy to break a paintbrush by accident—it was definitely on purpose.

FRIENDSHIP RECIPES

MIX AND BAKE
AND YOU'LL HAVE INSTANT FRIENDS!

RECIPE B: How middle-school kids think
friendships work—it's more complicated than
a math problem or a secret code.

- Add one part how you look.

- Stir in a cup of what you say.

- Then add a pinch of where you live.

- Toss in what your parents do and
  what kind of car they drive. Don't
  forget to add what clothes you wear.
  Somehow shoes are extra important.

- Mix well and maybe, just maybe,
  you'll have a friend.

FRIENDSHIP IN MIDDLE SCHOOL IS COMPLICATED

Who is a good friend? How do you be a good friend?

The rules for being a good friend are definitely different in middle school than they were in elementary school.

This used to be an easy recipe—or equation. Being generous plus being friendly plus being fun to be around equals a friend.

These things don't matter anymore. Now people only want to be friends with popular kids. And popular doesn't mean being nice. It means all kinds of weird things.

FRIENDS IN ELEMENTARY SCHOOL

The kid who plays in the sandbox with you

The kid who shares lunch with you

Have one.

Thanks!

The kid who saves you a place in line

Over here!

Thanks!

POPULAR KIDS
(as best as I can define)

THE KID WHO'S AN "INFLUENCER,"

whatever that means

It means I'm bossy!

You should do whatever I do. Follow me!

You're wearing that?!

THE KID WITH THE RIGHT CLOTHES

I didn't know clothes could be "wrong" unless they were the wrong size, but it turns out there are so many ways clothes can be NOT right. It's just that I don't know what they are.

dorky sweatshirts

droopy socks

old man shoes

old lady shoes

THE KID WITH THE MOST "LIKES"

I thought being liked in person mattered more. NOT according to the TOP SECRET middle-school rules.

Eyes glued to phone, never looking up

OBSERVATION # 11

I FAIL IN <u>ALL</u> THOSE AREAS. ALL OF THEM.

But then, so does Dash. So do Skye and Luisa (OK, they do have some pretty cute clothes—I just don't know if they're the "right" ones). I thought that's why we all got along.

DASH

Generic shirt, pants, sneakers, everything. His parents are as clueless about clothes as mine are.

SKYE

Shoes like ballerinas wear so she always looks elegant no matter what else she's wearing. Earrings because she was allowed to get her ears pierced and I still can't.

LUISA

Tops that look simple and fancy at the same time. I can't explain it— they just don't look like something you find everywhere, the way my clothes do. And earrings, of course. She got her ears pierced as a baby!

DEDUCTION # 2

THERE ARE DIFFERENT KINDS OF FRIENDS
(BESIDES COOL AND NOT COOL). NOT JUST HOW CLOSE
YOU ARE TO THEM, BUT HOW THEY FIT INTO YOUR LIFE.

PLACE FRIENDS

BUS FRIENDS: Kids you wait in line with, sit with, talk with,
but only on the bus or at the bus stop. Anywhere else and
it's like you're invisible to them.

PE FRIENDS: These are a rare breed for me since I'm so bad at sports. Anyone who's willing to be nice to me—or at least not mean—is an automatic PE friend. PE friends are teammates, kids you joke around with and support during games. I never get that far since I'm the kid who's always picked last for any team.

LUNCH FRIENDS: These are always actual good friends. Because being willing to eat with you is a social statement in front of the whole school. Bonus points if they pack good lunches and share them.

NEIGHBOR FRIENDS: Kids who just happen to live near you. You may not do anything together, but you at least say hi when you see each other. Is this the kind of friend Dash will become?

This kind of friend you might have played hopscotch with on the sidewalk when you were little. Not in middle school!

THERE'S ONE THING THAT DASH AND I WILL ALWAYS HAVE IN COMMON. IT MIGHT BE THE WAY TO STAY SCHOOL FRIENDS!

MATH!

Yes, math, the subject most kids groan about. But Dash and I think numbers are fun. See how cute and cuddly 8 is, kind of like Santa, only without the beard and presents.

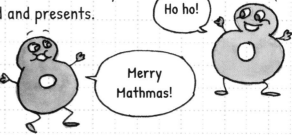

Ho ho!

Merry Mathmas!

Plus I like that math is SO clear. There are only right and wrong answers, no gray areas. If you follow all the logical steps correctly, there is that satisfying *C L I C K* in your brain when you solve a problem, like fitting a puzzle piece into place.

Clickety

Clackety

Where else in life do you get clean, straight answers like that? Not from your parents—"Because I said so" is the dumbest reason ever!

Friendly

Everyone's lucky choice

Keeps you company

Dash and I both tried out for the mathlete team, which meant we'd be together at every practice.

Mathlete = Math + Athlete. (Except no muscles needed, only brainpower. Good thing, too, considering how bad I am at PE. Math is the only "sport" I can be good at.)

A mathlete is someone who does math competitively. There's a timer and you have to solve the problem before the people from other teams do. Kind of like Jeopardy!, only with math, not trivia. Each team puts up one member for each question, so everyone gets to answer. You have to solve the problem on your own, not consulting with team-mates, so if you get it wrong, it's all on you.

BLUE TEAM

RED TEAM

YELLOW TEAM

GREEN TEAM

A competition could have four teams, each from a different school, like this. Even without running hard, mathletes sweat from thinking so hard (or being so nervous).

There can be a wide range of problems, but most are logic problems.

EXAMPLE:
Today is my birthday. If my age in months is 99 greater than my age in years, how old am I?

A) 9   B) 11   C) 12   D) 14

You can't tell just by looking at me or counting my freckles. You have to do the math!

ANSWER:
Turn the logic statement into an equation. My age in months is 12 times my age in years (since there are 12 months in a year), plus the extra 99. Only 9 works because 9 × 12 = 108 (months) and 9 + 99 = 108.

Here's the math: Make X your age in years. Then: X × 12 = X + 99 or 12X = X + 99. You need to move the Xs to only one side of the equation, so subtract X from both sides and you get 11X = 99. Now it's clear that 99 divided by 11 is 9! Easy-peasy!

See how neatly that all works out? It looks a lot trickier than it really is since you're given four ages and can always just try using each of them to see if the math adds up. In this case, the first answer, the first one you plug in to the equation, happens to be right.

Today Mr. Douglas called out the names of all the students who had tested well enough to be mathletes.

Dash jumped out of his chair, he was so excited. I held my breath, worrying, while a lot of other names were called.

It seemed like forever but then—finally!—I heard my name. I'm a mathlete!

IT TURNS OUT THAT DOING SOMETHING AT
THE SAME TIME IN THE SAME PLACE DOESN'T
NECESSARILY MEAN YOU'RE DOING IT TOGETHER.

Here's the uncomfortable, weird thing. At first I was really
excited to be picked, especially since Dash was, too, and that
meant we'd be mathletes together. Or so I thought.

The first mathlete meeting wasn't like that at all.

There was a wall of boys, including Dash, all sitting together.

I was the only girl in the room!

I know I'm a nerd. But I can't be
the only nerdy girl in 6th grade.
There have to be others!

I glanced at Dash but he wasn't
looking at me.

I tried to pretend I totally belonged and walked to the only empty seats, all in the back of the room.

Mr. Douglas introduced us to our team captain. If he thought it was strange to have only one girl mathlete, he didn't say so.

Welcome, mathletes!

This will be a great team and we have a great captain— Charles! He scored the highest on the math aptitude test this year.

$32(10 \div 8) - 267 = x$

This will be lots of work, but also lots of fun!

Mr. Douglas wrote problems on the board for us to do. I told myself, "I like math—this will be fun," trying to shut out everything else. And it worked for a while.

Because numbers are clear and they don't care if you're a girl.

BEING A MATHLETE ISN'T JUST ABOUT DOING MATH.

If it was, that would be easy. But it's also about being on a team. People who are good at math aren't necessarily good at working as a group—just the opposite, in fact.

I'm bad at team sports myself (well, any sport, really), always the last to be picked. At first I thought the math-lete team would be different since the reason nobody wants me on a team is that I'm so bad at sports. Can't blame anyone for avoiding a loser.

But I'm good at math.

Once again, I felt like the deadweight nobody wanted to have on the team. The other kids were laughing and joking together. Except for me. It was like I was invisible.

DEDUCTION # 4

MAYBE BEING A MATHLETE IS A MISTAKE.

I caught up with Dash walking home (after Echo Park, like he insisted). He was SO excited and happy, bubbling over with how great Charles is and how our team is really strong and will win the district championship for sure. He didn't even notice how quiet I was, which, believe me, is not normal. I wanted to say something sarcastic about how nice it was that he called it "our" team when he had totally ignored me.

I felt like an old toy that wasn't fun anymore.

No, he isn't. But I WANT to be friends with him. Is there something wrong with that? Friendship is a BAD thing to you now?

You should like him, too! This is your chance to prove that girls are OK and boys can be friends with them.

I didn't realize that was something that needed to be proved!

REALLY!?!

If Charles likes me, THEN you can like me? Like I have to pass some sort of test?

I could hear the whine in my voice but couldn't stop it. Who wants to be friends with a whiner?

Why should I have to make Charles like me? He should be the one reaching out to me since he's the team captain. Why is it <u>my</u> job to turn Charles into a friend?

And could it really be that simple anyway? Of course not! I'm terrible at making people like me. It's just not something I know how to do. Skye can do it. So can Luisa. They know how to seem like instant good friends. Not me. It's not that I'm mean. I'm a nice person, I really am. But that's not enough. Being smart is definitely not enough (in fact, it usually counts against you, makes people NOT like you). Whenever I try to get people to like me, it's a total failure.

OBSERVATION # 14

WORKING TO MAKE PEOPLE LIKE YOU RARELY WORKS.

There was that time at summer camp, when there was this girl I really liked and I desperately wanted her to like me back.

I love that birchbark canoe you made—SUPER cool! Plus your shorts! Super SUPER cool!! And your sandals, just amazing!!!!

Too eager, trying way too hard

Um . . .

BYE!

Can't get away fast enough

I wanted to stop talking, I really did. I could hear how stupid I sounded. "Amazing sandals"?!? But my mouth kept saying things, completely disconnected from my brain. She just stared at me and rushed off like I was crazy. And really, who could blame her?

What is that mysterious something that makes a person "likable"?

FRIENDSHIP QUIZ

Do you like someone who:

If you answered mostly A's, you don't want a friend—you want a servant.
If you answered mostly B's, you like adventures and taking risks.
If you answered mostly C's, you're not sure what—or who—you like.

I can't tell what kind of friend I am from this quiz. If I gave it to Charles, what would he say? Probably that it was a dumb quiz. He would want a different kind of quiz, something like:

FRIENDSHIP QUIZ

The most important thing in a friend is:

A) THEIR BRAIN	B) THEIR SENSE OF HUMOR	C) THEIR SENSE OF ADVENTURE	D) HOW NICE THEY ARE
I want Einstein for a friend	I want a clown!	I want a daredevil!	I want Santa.

The thing that matters least in a friend is:

A) THEIR LOOKS	B) IF THEY'RE GENEROUS	C) HOW NICE THEY ARE	D) HOW FUN THEY ARE
Not cute.	Selfish	Not nice.	Not fun.

If I have to make Charles like me—

I'M DOOMED!

Really, I should think of this like a logic problem. Then maybe I can find a solution.

If Charles thinks I'm good at math, then

    A) he'll respect me.
    B) he'll think I'm a valuable member of the team.
    C) it won't matter that I'm a girl.

So say Charles being my friend is X in the equation and Dash being my friend is Y. If X, then Y. Meaning if X is true, then Y is true. I need X to come true!

But this isn't the kind of logic problem I can solve. I'm not figuring out what X or Y is. I'm trying to make a possibility into a fact. That's not math at all. It's wishful thinking.

If only I were better at making people like me . . . . if only I could magically impress Charles. If only . . .

I tried to explain all this to Skye and Luisa the next day before class started.

It did sound stupid when they put it like that. But Dash isn't just any boy.

How can they? I don't really understand myself.

A FRIEND SHOULD LIKE YOU THE WAY YOU ARE,
NOT THE WAY THEY WANT YOU TO BE.

How do you know someone's really a good friend? What stuff matters?

FRIENDSHIP PIE

Every ingredient counts.

They laugh at your jokes.

They don't lie to you.

They listen to your problems.

They like the same things you do.

They stand up for you.

They are different from you in ways that are cool and interesting.

They are fun to be with.

They help you.

OBSERVATION # 16

LITTLE BROTHERS AREN'T AS DUMB AS THEY SEEM.

Leo is only 7, but after school that day, even he noticed something was going on with Dash.

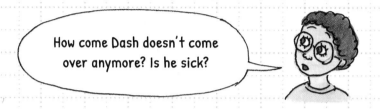

> How come Dash doesn't come over anymore? Is he sick?

My first impulse was to lie, to say that Dash was just busy. But then I thought, why lie? So I told Leo the truth.

> I guess we're not friends right now.

> That can happen?!!

> HOW ?!?!?!

Leo looked like he'd seen a ghost. I guess he's too young to have outgrown a friend yet.

Now I wish I had lied! This is all too complicated to explain to a little kid. I tried to tell him it was OK, people change sometimes, that's all. Which freaked him out even more.

You're going to CHANGE?!?

You'll stop liking me!?!?

Mom and Dad, too??!!!

WAAAAAH!

DEDUCTION # 5

I'M NOT ONLY A C MINUS FRIEND— I'M A TERRIBLE BIG SISTER.

How a big sister should be:

Shares her markers and sketchbooks with her messy little brother.

Funny, as in knows good knock-knock jokes.

Will always read yet another book to him.

Patient, as in puts up with question after question. (Why is the sky blue? Why do we cry tears? Why do knees bend forward and not backward?)

I tried to reassure Leo that family is different. I'll never stop loving him. He shouldn't worry about any of this—it's just stupid middle-school stuff.

I was just making things worse! Leo didn't calm down until I promised to play Life and Candyland and Uno with him.

At dinner that night, I stupidly told Mom and Dad about the mathlete team. I knew they would start grilling me. They might even start assigning me extra math homework.

Dad's an environmental scientist.

He's always reading some scientific journal or another, likes to putter in the garden and test out ways to control weeds and grow plants. Be careful about asking him questions because he'll give you a much longer, more complicated answer than you want to know.

Mom's a computer programmer, so pretty good at math herself. She plays computer games when she has free time, so she's basically always in front of a screen, either for work or for fun. She's not a touchy-feely, huggy type of mom like Luisa's or a *let-me-feed-you-something* mom like Dash's or a here's-a-great-book-you-should-read mom like Skye's. More of a don't-bother-me-with-boring-details mom. It's obvious where I get my nerdiness from.

I wasn't going to say anything. I figured I could rattle off a bunch of other names, talk about equations, just avoid the subject of Dash. But Leo piped up.

Suddenly he's the expert on how to lose a friend.

OBSERVATION # 17

THE WORST KIND OF STARE IS THE PARENTAL *YOU-WILL-TELL-ME-OR-I'LL-NAG-IT-OUT-OF-YOU!* STARE.

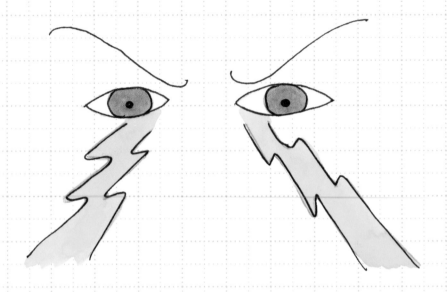

Mom and Dad both gave me that stare.

Mom's voice had that icky syrupy tone that moms are expert at, the *trying-too-hard* kind.

Leo could explain it all to them, not me.

OBSERVATION # 18

IF YOU LOSE A FRIEND, THE BEST THING TO DO IS
FOCUS ON THE FRIENDS YOU STILL HAVE. AND MAYBE
THEY CAN HELP YOU GET YOUR OLD FRIEND BACK.

Leo gave me an idea—I could show Dash that we can still be
friends. I just needed Skye and Luisa to help me.

I told Skye and Luisa today about the amazing idea I got
after the dinner fiasco last night.

These were not the encouraging faces I expected. They
didn't look thrilled, not one bit! I reminded them that they
loved the tech/coding camp they went to last summer. And
coding IS math. They didn't buy it.

Skye reminded me that to be a mathlete you have to be really good at math. She doesn't think she fits that description at all.

CODING SOUNDS COOL. MATH DOESN'T.

All the rest of the day, I tried to think of a way being a mathlete could feel more like coding camp, but I was coming up with a big, fat ZERO.

Then I had another idea—maybe there was a different kind of coding we could do as mathletes. People think of computers when you say coding, but I'm thinking of old-fashioned codes, the ones used to write secret messages. A code looks like gibberish but really means something super private or important.

CODING
PAST: How Spies Talk to Each Other
PRESENT: How People Talk to Computers

Grandma says that when she was a kid, you might find secret decoder rings in a box of cereal.

Nowadays all you find in a box of cereal is cereal.

Decoder ring with one alphabet on the outer circle, another on the inner one (or numbers).

In those days, the rings were a good gimmick to sell cereal.

Old-fashioned coding sounds like a LOT more fun than computer coding. Skye and Luisa won't be able to resist! But how do I make it part of mathlete practice? I could already tell that Mr. Douglas wanted to focus on the kinds of problems we'd face at competitions.

There was only one person who could help me with this:

DASH

Would he talk to me about this as long as we're away from school? Was this something we could do together (like old times)? Would he help me figure out secret codes that the mathlete team might be interested in? Could this be a way for me to prove to the whole team I can contribute something?

After all, codes could be a way to do something with math that's not competition. Just pure math fun we could all do together! Having more eyes on a code increases your chances of cracking it—it's a great chance for teamwork!

I worried I was putting too many hopes into one secret code basket. I wanted codes to make Dash be my friend again, to convince Skye and Luisa that being a mathlete can be fun, to show the mathletes what a good teammate I am.

That's asking a lot from a secret code!

I can do it!

IF YOU WANT TO SELL AN IDEA, YOU NEED TO PRESENT IT IN A CLEVER WAY.

Make it look like so much fun, nobody can say no (kind of like getting people to buy cereal by hiding a cool surprise inside the box).

I had one chance to sell Dash on helping me with the coding idea. So how to do it?

IN CODE, OF COURSE!

Me, in my spy disguise

with my hair up and fake glasses attached to a plastic nose

Obviously still me, but funny!

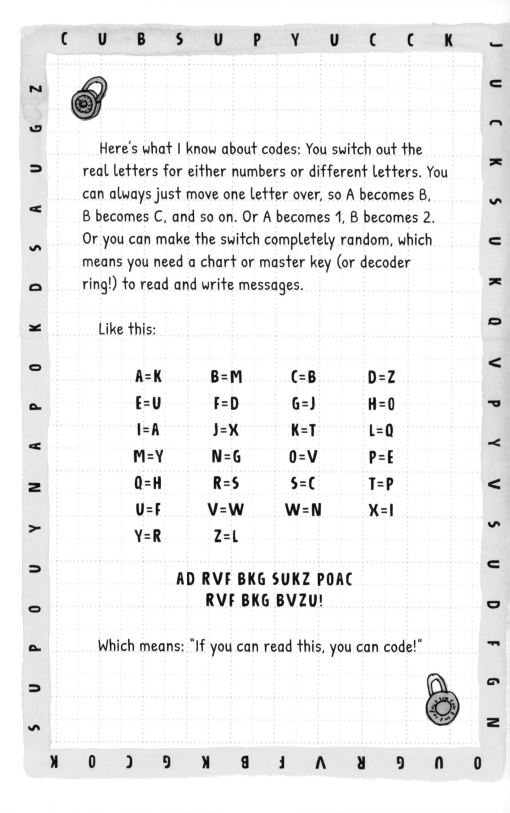

Here's what I know about codes: You switch out the real letters for either numbers or different letters. You can always just move one letter over, so A becomes B, B becomes C, and so on. Or A becomes 1, B becomes 2. Or you can make the switch completely random, which means you need a chart or master key (or decoder ring!) to read and write messages.

Like this:

A=K	B=M	C=B	D=Z
E=U	F=D	G=J	H=O
I=A	J=X	K=T	L=Q
M=Y	N=G	O=V	P=E
Q=H	R=S	S=C	T=P
U=F	V=W	W=N	X=I
Y=R	Z=L		

**AD RVF BKG SUKZ POAC
RVF BKG BVZU!**

Which means: "If you can read this, you can code!"

RAIL FENCE CODE

You don't switch out the letters for this one. Instead you rearrange the order, so it looks like nonsense instead of a message. First you write the message so every other letter is staggered below (like a fence), reading down, then up, down, then up in a zigzag pattern.

D   S   S   O   L   B   M   F   I   N

  A   H   H   U   D   E   Y   R   E   D

The message has to add up to four-letter chunks, so if the total letter count in the message isn't a multiple of four (like 20, 24, 32, etc.), then you add extra "null" letters. The null letters are placeholders, not part of the actual message, so it's good to use a letter that's not used often, like Z. My message is: "Dash should be my friend." That's already 20 letters, so no null letters are needed.

Now copy out just the top row:

DSSOLBMFIN

Add the bottom row after the top row:

DSSOLBMFINAHHUDEYRED

Looks like babble, right? But we can make it even trickier by chopping it up into 4-letter word groups (this is where the multiple of 4 thing comes in):

DSSO LBMF INAH HUDE YRED

If you got this message, you would have no idea how to read it, how to decode it. UNLESS you know it's in rail fence code. Then we just reverse engineer the steps for encoding to decode.

First put all the letters into one long string:

DSSOLBMFINAHHUDEYRED

Now put a slash halfway through, dividing the letters into two equal groups:

DSSOLBMFIN / AHHUDEYRED

Turn it back into a rail fence by putting the letters on the left on top, the ones on the right staggered below:

<pre>
D   S   S   O   L   B   M   F   I   N
  A   H   H   U   D   E   Y   R   E   D
</pre>

Now you can read it, going down and up in the zigzag:

DASH SHOULD BE MY FRIEND

It's true that there aren't breaks between words, but your brain will do that for you. Just like you see shapes in clouds, you see words in strings of letters.

That cloud looks like a sheep. Our brains like to make sense of things.

If Dash reads this code, will he want to be my friend again or does the message sound too desperate? That's the real code I need to figure out—how to talk to people!!

Leo came into my room (without knocking, which is a polite code for asking permission to enter), and before I could stop him, he grabbed my code about Dash.

I thought the rail fence might be tricky for Leo, so I showed him a different code. That shut him up.

THE KEY GRID CODE

Any five letters are your key. Inside the grid, you fill in the alphabet in random order. Since 5 × 5 = 25 (more math!), you're short one space to fit the whole alphabet. You can use "I" for both "I" and "J" or for both "I" and "Y." I'll use LUISA for my key:

	L	U	I	S	A
L	K	A	Z	N	W
U	F	I	Q	R	E
I	B	P	U	T	M
S	H	S	X	C	G
A	Y	L	O	V	D

To get the encoded letters, you use pairs from the key, first the side, then the top, so LL = K, LU = A, LI = Z, LS = N, etc. My message for Dash could be:

SA UU US AU SU SS LU LS SS AI AA UA

To make it trickier, I can put the letters in groups of five, so nobody guesses it's a key grid code. Then it would look like:

SAUUU SAUSU SSLUL SSSAI AAUAJ

The "J" at the end is a null letter, since the grid has no "J" and I need another letter to make a block of five.

I didn't know I was so mysterious!

Sometimes Leo really surprises me. Yeah, what he said was true. But it wasn't helpful.

How can I impress Dash if Leo keeps distracting me?

If Dash can't figure out the code, he'll definitely want me to explain it to him. He'll have to talk to me.

SOMETIMES HARD STUFF IS FUN IN A CHALLENGING WAY.
SOMETIMES HARD STUFF IS JUST HARD—AND FRUSTRATING.

The next day, I showed the code to Dash as soon as I could,
at the usual past-the-park place on the way home from
school. I thought he'd be amazed by how clever I am. He'd
see what a fun friend I can be, not boring at all.
    That's not what happened.

Dash didn't used to be so critical. But then he must have felt bad because he offered to come over and do math homework together.

For an hour or so, it was like it used to be. I remembered why I like Dash so much—he has such a different way of thinking about things than I do. Talking to him is like visiting a magical place where you always see something new. I love his descriptions of our teachers.

Doesn't Mr. Schaeffer remind you of a butler from an old movie? He stands so straight, it's like he's frozen in place.

And Mr. Douglas is like that guy who advertises toothpaste, with the extra-big smile all the time.

I knew he reminded me of something!

Plus math, of course.

What do you have for problem 16?

Oh, that's a fun one! You have to basically solve it backward.

Now I get it!

I told him the story about the girl with the disapproving eyebrows.

I hadn't realized that Dash had his own stuff to deal with, that he was getting picked on.

To me, he's perfect. I guess I thought everyone else saw him that way, too.

I'm sorry middle school is like this! So much has changed and I don't understand why.

So you'll be patient with me now?

I told him I would. I get how he feels. I didn't even bring up the idea of doing coding with the mathletes. It didn't seem like the right time. Dash just wanted to fit in.

Dash looked so relieved, I promised myself to at least try to do what he wanted. I'll even try to make Charles like me, though that seems pretty impossible. Still, I can at least show Dash I'm working on it.

SOMETIMES LITTLE BROTHERS CAN BE BIG PAINS!

Leo got home from soccer practice and barged into my room (again without knocking). When he saw Dash, he froze.

DASH????!!!

IT'S REALLY YOU!?!

Dash laughed.

Who else would it be?

I glared at Leo, trying to make it clear that he'd better not say anything stupid.

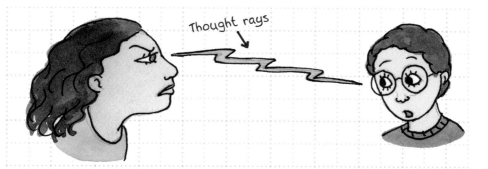

DON'T SAY ANYTHING ABOUT DASH NOT BEING MY FRIEND!

I had to say something fast, before Leo spilled the beans.
Something! Anything!

IF YOU WANT TO GET LEO'S ATTENTION, TALK ABOUT DINOSAURS.

Leo is obsessed with dinosaurs.

He knows every species and when they lived. He wants to be a paleontologist when he grows up, digging for dinosaurs in Utah or Arizona. Last year we went to Disneyland, but Leo had way more fun at the La Brea Tar Pits.

Nothing fascinates him more than giant skeletons of monsters that actually existed.

It's sad to think dinosaurs got stuck in tar and died.

But lucky for us that they did so we can see them now.

My distraction worked. Leo ran to get his T. rex while Dash packed up his stuff. Leo only had time to shove the model in Dash's face before Dash was out the door.

I'd given up on the code idea, but just as Dash walked out the door, he turned and said, "Think of a better code."

Sure? Sure! This felt like a big step back to Dash being my everyday friend, not my secret-only-sometimes friend. We have something to work on together!

EVEN PEOPLE YOU KNOW WELL CAN SURPRISE YOU.

I called Skye first, then Luisa, giving them the great news—Dash is still my friend (just not at school for reasons I get) and we're working on a code together. Skye said she wanted to help, too. So does Luisa. So now ALL my friends are in this with me. Even if we can't figure out a good code, we'll have fun trying.

The next day was mathlete practice again. I knew Dash wouldn't sit next to me, but now that I understand better why, it didn't feel so bad. And when I caught his eye, he gave me a quick wink. That made me feel like he was right beside me.

When Charles told us to do logic problem drills, I saw my chance to be a team player. I put up my hand.

How about playing with some codes, the secret language kind? Those help with logical thinking. We could work together, in groups of two or three.

We have a competition to prepare for! We're doing math. We're mathletes, get it?

I got it all right. I looked over at Dash. He was frowning. Everyone hunkered down and started solving the problems as Mr. Douglas read them out. I tried to focus only on the math, not think about Dash or Charles.

Talia, what's your answer to the first problem?

Um, I haven't solved that one yet. I skipped it to work on the second problem.

That strategy works on a written test, but not in a mathlete competition. You have to solve the problems as you're given them.

He could have told us that!

Three boys behind me started snickering, and I could feel my face go hot.

I didn't hear the answer Ira rattled off. I was feeling too rattled myself. It was just a stupid mistake, I told myself. I WILL do better next time.

I shuffled out of the classroom when practice was over. Would Dash want to walk home with such a loser? I didn't think so.

But after Echo Park, there he was.

Don't take it so hard. Mr. Douglas is trying to help.

And I made the same mistake.

You did?

Yeah, we're not used to competition. It's all new for us.

But those boys were laughing at me!

Dash laughed himself and said now I know how he feels. He had a point, but I hadn't even gotten a chance to talk about my coding idea. I explained how solving secret codes, the kind spies use, is a fun way to work on our logical thinking, plus work together as a team. Codes are part math, part pattern recognition, and all teamwork, since each person brings fresh eyes to the problem.

Dash wasn't convinced. Not yet at least. But I'll work on him. I know I'm right about this!

MAYBE MIDDLE SCHOOL IS A CODE I CAN CRACK!

For example, how do you understand sarcasm?

THE LANGUAGE CODE

WHAT SOMEONE SAYS:	WHAT IT MEANS:
That cafeteria meatloaf smells delicious!	How can you eat that? I'm gonna barf.
I'll text you later.	You'll never hear from me.
Where did you get those clothes?	Throw them out, now!
Bet you're doing something fun this weekend.	If doing homework is your idea of fun.

The trick with the language code is that so much is said with the tone of voice, more than the words themselves. So if you're tone-deaf, you'll never understand anything! How can you tell when "I love it!" actually means the exact opposite, or "Only a complete idiot could possibly like that!"?

Getting the exact right snarky tone is a real achievement. Sadie, a girl in my English class, is the master of it.

Really?

Reeeeeeeeeally?!

THE DRESS CODE

(and I don't mean what the school says you should or shouldn't wear)

COOL

(which can change from month to month, even day to day,
making it extra hard to figure out)

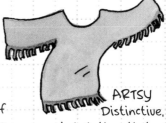

POPULAR
The expensive shoes the cool kids all wear.

ATHLETIC
The cap is cool only if it's for the right team. Bonus points if you leave the price tag on for some mysterious reason.

ARTSY
Distinctive, weird clothes that only an artist type can pull off.

NOT COOL

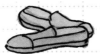

NERDY
The kinds of clothes (knit shirts with collars, for example) that only nerds wear.

FRUMPY
Loafers are never cool.

FREAKY
Oddball hand-me-downs from your mom that are super embarrassing.

DEFINITELY NOT COOL

(or everything I wear)

old lady socks

generic shoes

homemade shirts with felt appliqués stitched on (my mom really made these!)

THE WHERE YOU SIT AT LUNCH CODE

ONCE YOU START LOOKING FOR CODES, YOU SEE THEM EVERYWHERE.

By that I mean hidden meanings. They're in advertising, in the way people talk and dress, in the way we're graded at school.

When I walked to school, I told Dash about all the non-math codes, the ones people use to "decode" or understand a social situation. I showed him my middle-school codes examples. We just need to crack the weird code about girls and boys not being friends for some reason now.

EXAMPLE #1: Girl being mean.

EXAMPLE #2: Boy being mean.

There may be a code in the insults ("fart face" and "tuna breath" both stand for "stupid jerk"), but the meaning is clear.

Dash says texting is also full of codes. And I'm also getting them all wrong. Like <3. What does that mean?

A witch mooning you?

A double ice cream cone getting spilled?

A chicken?

A wizard with a double chin?

A man with a mustache wearing a pointy hat?

THE ANSWER IS A HEART!

Who figured that one out?

This explains a LOT about why I'm so bad at texting.

Another big middle-school code is just in your face—like the eye roll and the nostril flare, the kind the girl gave me when I sat next to her.

FACIAL CODES

Eye rolls mean many different things, but they all basically say that whoever is doing the eye roll is much smarter than whoever or whatever they're rolling their eyes at. Maybe not exactly "smarter," but better.

Stupid!   Get a load of this!   Not this again!

Are you insane?   I don't believe it!

Did you know that even dogs roll their eyes? When they do it, it means they want a treat. Or they've chewed up your favorite shoes and they hope you won't find out.

Here's another elaborate code.

HAIR CODES

MESSY	PERFECTLY ARRANGED	PLAYFUL	SERIOUS

| I'm too cool (or too totally NOT cool) to care about my hair. | I'm so cool, even my hair is cool. | Even my hair has fun being on my head. | I like the disciplined military look, no fooling around. |

Even after I think I've cracked the code to kid hair, adult hair—teacher hair—totally confuses me. Why is a man bun cool but a woman bun the opposite of cool?

MR. LOPEZ, THE PE TEACHER	MS. BLOCH, THE LIBRARIAN

Hipster dude with carefully managed hair—from mustache to beard to man bun.

Anti-hipster—in fact, a kind of cliché. In stories, librarians always have glasses and buns and are frumpy. Is Ms. Bloch dressing the part?

Which makes me realize that stereotypes are another kind of code—shortcuts for reading people.

I'm not sure what my hair says about me except that it's untamable. Nobody can boss my hair around!

I was guessing she didn't mean a math product, like the product of 2 × 2 is 4.

Dash liked my middle-school codes, but he still didn't want me to suggest secret codes for the mathlete team. He said it would be better to wait until after our first competition. There was too much pressure to win now.

He had a point. And I could wait. The important thing was that now he agreed that codes are fun and a good way to sharpen our brains—as a team.

BEING A MATHLETE DOESN'T WIN YOU ANY POPULARITY CONTESTS
(except with parents, and they don't count).

I thought Luisa and Skye would be proud of me for deciding to stick with the mathlete team. But when I told them, they acted like there was never a chance I'd leave.

Plus they knew I wanted to work on the code idea. And it turns out they did some research.

Code idea from Skye: **HIEROGLYPHICS**

We can use ancient Egyptian writing! It's both symbols and phonetic. Like a leg means a leg but also the letter B. Isn't that cool?

Picture writing is totally cool! It's kind of like a rebus.

  a

What a great eyedeer (idea)!

 = A or vulture   = C or cup   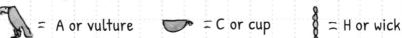 = H or wick

According to Luisa, hieroglyphics are like the original emojis, except we use emojis for tone, not information. Tone is another code! For example, you can text "You did it!" and if you add:

 It's an accusation.

 It's a celebration (that's supposed to be confetti).

 It's a surprise (though this emoji always looks a little worried to me).

If you don't have an emoji, you don't really know how to read the text.

Luisa had a code idea, too.

I like the idea of a decoder ring like your grandma talked about, except that means everyone needs to have one. Better to have a code where you can just tell people how to solve it.

THE GRID CODE

It's kind of like the rail fence code but even easier to solve if you know what to do. Here's how it works. You send a message that looks like complete nonsense. For example:

Y H N O I I O A O R S S U V W E A C J E H A N L U T O D D E S O W T I A T K T H T R

It looks like babble, but if you know that the solution is to put these letters into a seven-by-six grid, starting from the upper left corner and moving down, then you get this:

```
Y  O  U  J  U  S  T
H  A  V  E  T  O  K
N  O  W  H  O  W  T
O  R  E  A  D  T  H
I  S  A  N  D  I  T
I  S  C  L  E  A  R
```

Now you read it from left to right, top to bottom, and it all makes sense. Pretty cool!

I told Luisa and Skye that this is proof they should be mathletes. They still aren't convinced. But I LOVE their codes. I can't wait to share them with the mathletes—once we win the competition.

On Monday we finally found out when exactly that will happen.

Our first contest is in three weeks, competing against other schools in the district, so we need to meet every day after school to prepare. Starting today!

How do we "prepare" for a competition like that? You can't sharpen your brain like you can a pencil. I dreaded spending every afternoon with a bunch of boys who are judging me, waiting for me to fail, but I told myself I'm not giving up. Even as all eyes glare at me every time I come into the room.

The worst is that Mr. Douglas leads the glares, and he's

the one who picked me for the team. Why, if he's going to be so hard on me? I guess I failed him somehow. I'm not as good as he thought I would be.

OBSERVATION # 27

"PREPARE" TURNS OUT TO BE A NICE WORD FOR STUDY, STUDY, STUDY. (AND TAKE A LOT OF PRACTICE TESTS SO WE CAN BE EXPERTS AT SOLVING LOGIC PROBLEMS.)

At mathlete practice, Dash still won't sit next to me. But he passed me this note:

"Forget about the code for now. Just study for the mathlete meet. You HAVE to get a good score."

That means he's rooting for me.

I like taking tests (and yes, I know that's weird). I can feel my brain racing and my heart pounding. It's exciting and scary, all mixed together. Kind of like a mental roller coaster. (Not an actual roller coaster, which I hate.)

Some people like horror movies. I like tests. It feels like the cogs and levers in my brain are clicking, turning, moving, figuring everything out.

Brain going on a wild ride. ↗

TIME!

Pencils down!

I felt like I'd just run a mile. But I solved every problem—correctly, I hoped. Charles read out the answers. I did pretty good, only three wrong. Dash got four wrong. (I admit it felt great to beat his score.) One kid got seven wrong, and I couldn't help but feel sorry for the way Charles glared at him. So maybe I'm not the worst mathlete after all, Charles! Just as I was feeling bad for the boy with so many wrong answers, Charles turned his glare on me.

I only got three wrong!

Why are you looking at me?

And we should applaud you for that? How about you think about your fellow mathletes for once and give them some help.

I don't see you ever offering suggestions.

That made me mad! So even though I'd promised to wait until after the competition, I had to say something.

I was fuming, gathering my stuff as all the boys snickered. Well, not all. But a lot of them.

Dash clearly felt bad because he walked home with me, right from school.

Even if I got all the prob- lems right, Charles wouldn't respect me. I can't just be good—I have to be the best. And then I also have to make everyone else better!

I think he was mad because you didn't bomb out, like he thought you would.

What?

He wants me to fail?!

Just because of that first mistake? Or because I'm a girl?

Dash looked all embarrassed, so I KNEW there was something he wasn't telling me. Something even worse.

What is it? Spill it! NOW!

No, nothing.

You're such a bad liar! Your ears turn bright red whenever you lie. So tell me what Charles said. TELL ME!

I was shocked!

I was afraid
I would cry!

And then I was
FURIOUS!

That did it—NO WAY was I quitting. I'd suffer through a million mathlete competitions just to prove those stupid boys wrong.

Why do they think I can't cut it? No, don't answer that—it's because I'm a girl. What rule says girls can't like math, can't be good at it?

How else was I supposed to put it? Then I asked him how I was supposed to fit in when everyone treated me like a big fat ZERO. I thought I'd get some sympathy.

That's not what happened. Dash thinks that I'm the one staying apart from everyone else. He admits some boys laughed at me, but says I'm the one who sits in the corner and doesn't say hi to anyone.

I bit my lip but didn't say that nobody says hi to *me*. Obviously, I'm supposed to be the friendly one!

Dash said Charles expected me to help out by talking about ways to solve problems. Suggesting we work on codes wasn't what he wanted. He wanted math solutions.

He made it sound like it was my fault nobody would sit next to me. I asked Dash what I was supposed to do. It felt like an impossible situation.

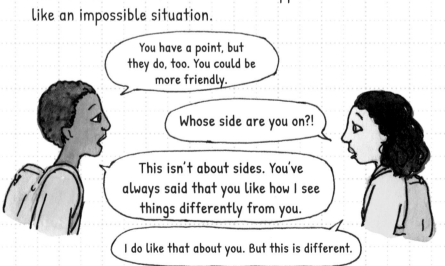

Exactly—it's different! You stay away from everyone, like it's OUR job to make you feel welcome when really it's your job to be a good team member. You have to fit in with the mathletes, not the other way around.

So the oddball had to fit in with the majority? That's what it felt like to me, but Dash insisted it wasn't about that. I could be the only girl. Did I just have to be friendlier? I wondered. I had tried, hadn't I? And they weren't exactly welcoming to me.

It's not so simple.

Why not?

You make them feel bad that you're there. Like you're judging them or something.

So now I'm not a team player?

Maybe not.

Dash is wrong. I can be a team player—with the right team.

That gave me a brilliant idea, the best one of all my recent ideas—I'm going to start my OWN mathlete team, ALL GIRLS! Maybe Luisa and Skye won't join, but I bet I can find others who will, girls who just need a little encouragement.

Dash's way of apologizing:

Um, do you want to study together?

Nope!

You gave me a great idea and I need to work on it.

Can't explain right now, but thanks for your honesty. You're a real friend.

And I meant it.

SOMETIMES WHAT LOOKS LIKE A WALL IS REALLY A GATE. YOU JUST HAVE TO FIGURE OUT HOW TO GET THROUGH IT.

The official mathlete team might have a big barrier against girls, but they're inspiring me to break down the wall. I'll show them who can stick it out in tough competitions.

I want to have our first girls' mathlete team meeting at our house, and I know I'll have to bribe people to come. Which means I need full parental support, even if it comes with parental nagging.

Mom and Dad were thrilled. They think my team will look impressive when I apply to colleges. (No pressure, of course.)

I told Leo he's too little, but he can start training with us so by the time he's old enough, he can be the mathlete captain and do a much better job than Charles, starting by being nice to girls.

I'm always nice to girls! Aren't I the best brother ever?

Next, I had to explain my big idea to Skye and Luisa.

SOME PEOPLE JUST REALLY DO NOT LIKE MATH.
I DON'T GET IT, BUT IT'S TRUE.

I tried to convince Luisa and Skye to join my new team—they saw how much fun codes are. I stuffed so many happy celebratory emojis into my texts that they would have to agree.

The emojis did part of the job. Luisa and Skye love what I'm doing but absolutely, no question about it, refused. Luisa said she totally supported my idea, but she couldn't possibly compete at math.

I'm just not competitive enough. Or mathy enough. 🙁

The first part's not true—she's super competitive. Try playing any game with her, even Monopoly. But the second part probably is.

Skye said the same thing. Instead she offered to find other girls to join the team. Luisa will help, too. So tomorrow, we're putting out the word: Girls Can Do the Math!

MIDDLE SCHOOL IS MUCH EASIER
WHEN YOU HAVE GOOD FRIENDS WHO SUPPORT YOU.

Being popular doesn't matter. Maybe fitting in doesn't really matter, either. All you need is one really good friend, two if you're lucky, three if you're super lucky.

Luisa and Skye weren't kidding. They put this poster up all over school:

Go, team, go!
GIRLS ONLY

SHOW WHAT
YOU CAN ADD UP TO!

PIZZA MATHLETE
MEETING

Talia's house
Friday night!
Don't miss it!!!

I didn't believe them. We still had work to do. The next few days, we talked as loudly as we could about how much fun math competitions are.

No uniform required.

No sweating.

Useful skills—knowing math saves money!

Fun puzzle solving!

Big feeling of achievement and power with zero muscle aches!

Did I mention fun? Numbers are your friends!

Dad wanted to know how many kids to expect on Friday night. Mom wanted to know how many pizzas to order. I had no idea, so I repeated what Skye and Luisa had said: "A lot!"

I'm still going to the mathlete practices—I definitely don't want anyone to think I'm a quitter. But I'm just taking notes for what to do with my new all-girl team.

You guys are getting better, but there are still too many sloppy mistakes.

No shortcuts with math! And that means you, Talia.

Shortcuts are my favorite—and usually they work for me. Why plow through a whole problem when you can take an educated guess? But I know better than to say that.

Of course, Charles. Do you want me to show you my work? Or shall I explain it to the group?

Another time maybe. Just get the right answer.

The right way.

So much for helping explain things to teammates. I tried!

Another great session, everyone. See you tomorrow!

And now I loudly say, "Hi, everyone!" and "Bye, everyone!" when I come and go.

Hello, all!

You see how you can solve this problem?

Good work, guys!

Thanks for the tip!

That's the team spirit!

Not that anyone says hi or bye back. Not even Dash.

FRIDAY NIGHT

Finally it was Friday. Luisa and Skye came over to be good friends (and to have pizza with me). We waited nervously. And waited. And waited. The clock was crawling like it does at the end of school.

DING   **DONG**

Then the doorbell rang. For 3 minutes and 36 seconds it rang—again and again and again and again (you get the idea).

Leticia was first. Then Clara and Lucy and Surya and Rhonda . . .

Until there were 16 of us (plus Luisa and Skye)! It was amazing! We started by introducing ourselves and saying what we like about math.

I love that math describes nature, like the golden rectangle and Fibonacci numbers.

I love that there are real numbers and imaginary numbers!

I love logic games and puzzles—they're so much fun!
I love that math always makes sense.

I love that there are clear answers—2 + 2 is never 6!

I love that math describes the biggest things and the smallest.

I was ready with some problems to solve (ones I copied from mathlete practice) and everyone did great. Then we had a big discussion over what to call ourselves. I wanted the Mathlete Dragons because I love drawing dragons. Plus I know how much Skye likes them.

We're powerful and magical. Who doesn't like us?

Leticia wanted the Mathlete Mermaids. I thought that sounded princessy, but she gave a whole long explanation that was actually cool. A long time ago, there was this tough math problem about how you should be able to use math to describe the rotation of an object (like a spinning top or a spiraling football). Men had been trying to solve it for a really long time, but nobody could figure it out. The solution seemed so impossible, they called it the Mathematical Mermaid, since mermaids don't really exist, just like the solution to the problem.

Everyone was hanging on to every word Leticia said, like she was a magical creature herself.

She explained that the woman was Sophie Kowalevski, a Russian mathematician from the nineteenth century. She lived at a time when women definitely didn't do math, but she didn't let that stop her. She not only solved a tough math problem that men had failed at; she was one of the first women professors of mathematics.

How could a dragon compete with that? So now we're the Mathlete Mermaids, because, like Leticia says, believe it or not, girls who are good at math DO exist!

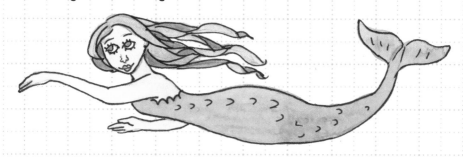

Luisa and Skye were a big help with organizing everything, so we're making them honorary Mathlete Mermaids. Meaning, they're on the team but don't have to do any math.

They already had clipboards with everyone's name and phone number. They moved to finding a sponsor right away.

I thought we should pick a woman teacher, but the teacher who got the most votes was Mr. Wayne, the art teacher. Art?

There can be a lot of math in art, like in fractals. Plus Mr. Wayne is a nice, fun person.

Yeah!

Exactly!

A repeating pattern is a fractal. There are fractals in nature, too. I'm already learning something new from my fellow Mermaids!

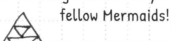

We finished up with homemade brownies (thanks, Dad!) and a big sense of accomplishment. I've never started my own club. I've never had so many friends. Now, suddenly, I have 16 new ones!

TOO MANY BROWNIES AND TOO MUCH EXCITEMENT
MAKE IT VERY HARD TO GET LEO TO SLEEP.

But I couldn't blame him. I was excited, too. I told him to calm down by counting sheep—see how useful math can be?

After school on Monday, I didn't go to the regular mathlete meeting. Instead I went to the art room to talk to Mr. Wayne. Everyone said he's friendly and easy-going, never mad or grouchy. But would he want to be our sponsor? I couldn't figure out a good way to ask, so I just blurted it out.

Then I realized he might not know what a mathlete team was, so I explained it to him.

Plus there's math in art, like in patterns and fractals.

I tried to remember anything else Leticia had said, anything that could convince Mr. Wayne. But I was drawing a blank (no pun intended!).

Fractals, eh?

Sure, I'll do it.

I guess you want me to file all the forms with the principal, which I'm happy to do.

Forms? Principal? I had no idea.

SOMETIMES WORRYING ABOUT DOING SOMETHING IS WORSE THAN ACTUALLY DOING IT.

He said yes! I convinced him! Maybe being a mathlete was giving me some kind of superpower for real!

Thank you, Mr. Wayne! We won't let you down. We're going to win our first competition, I promise. So can we meet tomorrow after school?

Sure thing.

Here in the art room. And let everyone know they should plan on a little drawing along with math.

Right away, I rushed to tell Luisa and Skye the good news. I wanted them to come to the meeting, too. They didn't say for sure they would, but I bet they do, if only to cheer us on—and do a little drawing. Who can resist an art room?

I was walking home in a happy daze when Dash ran up to me.

Hey, where were you? You missed the practice.

I thought you weren't giving up.

I'm not!

Dash should know me better than that! I'm not a quitter, not now, not ever.

But, then . . .

I was busy today. Something came up.

I didn't want to tell Dash my news yet, but he had a point. I had to go to one more mathlete meeting so I could officially resign. Resign from their team, not quit being a mathlete.

IT'S IMPOSSIBLE TO BE IN TWO DIFFERENT PLACES AT ONCE.

Unless, of course, you can bend the space-time continuum, which I can't. So I'll start at the Mathlete Mermaid meeting tomorrow, then run to Charles's team practice to quickly resign, then hurry back to study with my new, improved team. (Funny, but I'm actually excited about the drawing part.) I can't wait to see how those boys react. I'll show them who's a quitter!

THE ART ROOM

Skye and Luisa came with me to our first official Mermaid meeting. Yay! I asked them to get things started for me while I tell Mr. Douglas I'm changing teams.

Welcome, everyone!

Let's get started!

Mr. Wayne actually began the meeting, which made sense. So I felt I could safely leave for a bit.

I ran to Mr. Douglas's class.

I shot him a glare, then turned to face Mr. Douglas.

I explained that I was starting a new mathlete team, an all-girl one, the Mathlete Mermaids. The look on Charles's face was priceless!

I told him that Mr. Wayne was taking care of everything, and as soon as I said it, I could see that Mr. Douglas didn't take us girls—or Mr. Wayne—seriously.

Some of the boys started snickering. Maybe we should have picked a math teacher sponsor after all.

I've never thought of myself as competitive. I'm stubborn about a lot of things, but when it comes to sports, I know how terrible I am and give up right away. Not this time. This time, I really want to win. I need to win. It's a strange feeling, almost like being mad. Is there a good kind of mad, the kind that makes you focus and try even harder?

I had to rush back to my real team, but I glanced at Dash before I left. He looked stunned, like I'd dropped a bomb or something. Will he miss me? I hope so!

When I got back to the art room, I thought Skye and Luisa would be running things for me. That's not what happened.

Leticia had taken over everything. I thought about fighting her, shoving her aside—for about a second. She's so much more organized than me. And smarter. And prettier. And actually cool! I'm just determined and stubborn, which helped get us started, but she's a natural leader. Plus after fighting with Charles, I didn't have it in me to fight anymore.

Part of me is annoyed with her. Part is jealous. But the biggest part wants her to like me, to be my friend. I couldn't get Charles to like me, but maybe I'll have better luck with Leticia.

OBSERVATION # 34

IT'S IMPORTANT TO KNOW
YOUR OWN STRENGTHS AND WEAKNESSES.

I may be good at a lot of things (well, some things), but I'm not a leader-type person. I keep telling myself it's actually kind of a relief that Leticia is taking charge. Now if we lose, it's not my fault. Not that we're going to lose. No way!

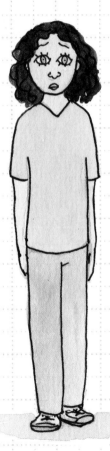

STRENGTHS:

I'm smart.

I'm stubborn.

I'm determined.

I can be funny.

I'm a loyal friend.

I try to listen to people.

WEAKNESSES:

I don't know how to talk to people unless I know them well.

I'm awkward in all kinds of stupid ways.

I'm bad at sports.

I have no fashion sense.

I'm good at organizing stuff for myself, but not for others.

STRENGTHS:

She's smart.

She knows how to talk to people and get them excited about things.

She is good at organizing (herself and others).

She is confident in herself. (I think that's what makes her popular.)

WEAKNESSES:

I can't see any, but there have got to be some! Maybe she's bossy or full of herself. Could be.

Leticia had us pair up and work on problems together, then present our solutions to the group. She says that teaching other people helps us learn ourselves. I guess that's what Charles was trying to tell me in his annoying way. But that's also what working on codes could do! I just have to figure out a convincing way to talk about them.

She's an inspiring cheerleader. Charles made us feel like we weren't good enough (at least he made me feel that way). Leticia has full confidence in us.

She has everybody energized and excited. By the end of the session, even Luisa and Skye wanted to be mathletes. I couldn't believe it!

She makes it sound like so much fun!

Plus all these girls are great! And guess what? Turns out we're better at math than we thought.

I've been telling you that for years! How come you didn't believe ME?!

Well, you know . . .

Actually, I do know—Leticia is cool. I'm not.

But I can't complain because now they really want to study math with me. And they've got some catching up to do. I wonder if I should still study with Dash. Or is that like helping the enemy since now he's on a rival team?

I hadn't decided what to do about Dash when he caught up with me walking home after mathlete practice.

Hey, Talia, that's cool that you're on an all-girl team.

I had no idea there were that many girls who are good at math.

Well, now you know.

Maybe we'll even beat you.

We walked in silence for a little while. I think we were both wondering what this all meant for our friendship. Then Dash surprised me.

And Talia?

Yeah?

I loved the look on Charles's face when you told him about your new team. It was priceless!

That made me feel sure that we would find a way to be friends in school again.

Could this day get any better? I had a new all-girl mathlete team with a nice teacher sponsor, and Dash was starting to agree with me about the old team! I told him I was sorry he couldn't escape Charles and join the Mathlete Mermaids. Then he surprised me again.

Dash finally gets that Charles isn't such a great team captain, but Dash still likes his team.

I was wrong about Charles.

About bothering to impress him.

I'm sure your team is great, but I'm OK with our old team.

I've made friends with two of the boys there, and that makes a big difference.

I was so happy for him! We were both making new friends—all thanks to math!

YOU KNOW THE SAYING "YOU CAN'T HAVE TOO MUCH OF A GOOD THING"? WELL, IT'S NOT TRUE.

I like math, but after a week of studying, studying, studying, even with a bunch of new friends, I've really had enough. We've been cramming so much math into our heads, I feel like numbers will fall out of my mouth. I'll sneeze out multiplication signs.

$8 \times 16 (3 + 22) = x - 12$

How many triangles do 6 intersecting lines make?

$\sqrt{3,832} = ?$

Now Dash walks all the way to and from school with me. We're both too full of math facts to worry about what other kids might think. (Well, I never worried, but Dash did.) Plus now Dash is friends with his teammates Adrian and Leroy, so he doesn't feel like he has to be as careful (at least walking to/from school—school itself is still off-limits). Which is lucky because having Dash around again makes me feel better about how Leticia's taken over everything and how infuriatingly perfect she is.

Sometimes I think she cares too much. She starts each meeting with a little speech and a fun problem. She talks to each girl about how they can improve. She's a whirlwind of motion, always doing something. She never gets sick of math! I know I can get her to like codes—I just have to pick the right moment.

SOMETIMES THE BEST WAY TO PREPARE FOR SOMETHING IS TO TAKE A BREAK.

I had a great idea about how to have fun, keep my brain from exploding with math facts, and build up team spirit. I'm hosting a math party! Well, a math scavenger hunt. Well, really a puzzle scavenger hunt—part math, part codes. This is how I can introduce codes to the Mermaids— they'll see it's logical problem solving. Plus this is a fun way for me to get back some control over the Mathlete Mermaids. Leticia is clearly the team captain, no way I can change that. But I can still do something special for all of us, show that I'm an important team member.

I can be a powerful mermaid! And a team player!

Here's how the puzzles will work:

1) The girls start out with a riddle to solve.

2) The answer to the riddle leads them to a message written in code and some jigsaw puzzle pieces.

3) Once they crack the code, they know where to find the next coded message and more puzzle pieces.

4) They solve the new code and are directed to one last puzzle piece.

5) Everyone puts the puzzle together, using all the pieces. The result is a map, showing where to find the treasure.

6) They follow the map and find:

PIZZA TREASURE

Another math pizza party!

IF YOU WANT TO SELL YOUR PARENTS ON AN IDEA,
THE BEST WAY TO CONVINCE THEM IS
TO CLAIM IT'S EDUCATIONAL.

If you can include math or science, you'll be even more effective (at least with my parents). The logical deduction is that parents are suckers for anything they think will make their kids better students.

When I asked Mom and Dad if we could host another Mathlete Mermaid meeting at the house, this time part of a puzzle scavenger hunt, they said yes right away.

145

I just need to be sure my parents don't get too involved. If I'm not careful, they'll start tutoring everyone!

Leticia agreed that a party like this is just what our team needs after so much hard practice. But she's not taking over the party—she can't since it's at my house.

We're set for Saturday. There's a lot to do before then, starting with thinking of riddles and codes.

After an hour at my desk with not a single idea for a riddle or a code, I decided to go outside, into our backyard. Maybe the fresh air would help. It couldn't hurt.

I was staring at my blank notebook—still no ideas—when Dash looked over the fence. Did I mention he lives next door, which is why we walk to school together?

Hey, Talia, what are you doing?

I explained the whole thing and he loved my puzzle scavenger hunt idea.

I told him it sounded good in theory, but I was stuck. I couldn't think of anything. Maybe I broke my brain from too much studying.

Dash hopped the fence and together we came up with a riddle and some codes. My brain wasn't broken after all! Together, we drew up the treasure map and cut it into puzzle pieces. The trick was to make things not too easy, but not too hard, either. Dash was great at finding the right balance. It was fun and relaxing, everything I wanted it to be. I'd been wanting to work with Dash on codes for a long time, and now we were finally doing it!

SORE, SAGGY BRAIN

FRESH, ALERT BRAIN

CLUE #1: RIDDLE

The more you take away from me,
the bigger I get.

ANSWER:
A hole! The biggest hole in town is
famous.

Kids call it The Pit.

It's on the edge of Echo Park, where
an old well once was. Now there are
boards over it and bright yellow
"DANGER" tape warning everyone to
stay away.

We'll stick the next clue, the code,
along with the first puzzle pieces in
a baggie under a rock near The Pit.
Dash will stand guard on Saturday to
make sure nobody finds it before the
Mathlete Mermaids.

Stash of
secrets

CLUE #2: CODE

15 12 12 16   25 22 19 18 13 23
  7 19 22   8 22 22 8 26 4

HINTS TO CRACKING THE CODE:

Clearly each number stands for a letter. The most common letter in the English language is E. The most common number in the code is 22, so you can guess that 22 = E. You can guess that the three-letter word ending in E is probably "the." So in this code, 7 19 22 = T H E, meaning 7 = T, 19 = H, and 22 = E.

Now it's time to make a smart guess. There are 26 letters in the alphabet. If each number represents a letter, you might think that 1 = A, 2 = B, 3 = C, 4 = D, and 5 = E. But we know (or are pretty sure) that 22 = E. What if the numbers are backward? Then 26 = A, 25 = B, 24 = C, 23 = D, and 22 = E.

That matches our assumptions!

You've cracked the code now! You just have to make a chart, like so, and then read the coded message.

26	25	24	23	22	21	20	19	18
A	B	C	D	E	F	G	H	I
17	16	15	14	13	12	11	10	9
J	K	L	M	N	O	P	Q	R
8	7	6	5	4	3	2	1	
S	T	U	V	W	X	Y	Z	

The secret message is
LOOK BEHIND THE SEESAW.

In that same park, there's a seesaw by the swings.
We'll hide the next clue and puzzle pieces right near it.

CLUE #3: CODE

The next clue is a word jumble.
All the right letters are in each word, but they're all
mixed up, in the wrong order. You have to unscramble
them to figure out the correct words.

IBMLC OT HET PTO FO ETH DSLEI

The short words make this easy (and are a big hint
this code is a word jumble rather than a letter-
replacement code). Reading only the short words, you
can see it says:

_ _ _ _ _ TO THE TOP OF THE _ _ _ _ _

Again, the word "the" is the easiest one to figure out.
Now it's not hard to unscramble IBMLC
and DSLEI.

The answer is

Whee!

**CLIMB TO THE TOP
OF THE SLIDE.**

Once you slide down,
you'll find the next clue.

CLUE #4: CODE

I added one last code, the trickiest of all,
because I couldn't resist.

```
G  K  T  I  A  S  E  O  O  I  D  B  B  L  L  E  B  A  R
O  C  O  L  S  U  A  L  K  S  E  E  Q  I  I  H  A  Y  D
B  A  T  A  H  O  N  D  I  N  T  H  G  R  N  T  C  K  !
```

How do I untangle this?

Maybe just chew through it?

I showed this one to Dash to see if he could solve it.

Is this some random code where I need to know the key, like the one you made before?

Because that's not fair—it's too frustrating for a party.

Nope, this one is much easier.

Dash stared at the letters for a long time.

It doesn't look easier.

And since you don't have spaces indicating words, I can't even look for "a," "to," or "the."

That makes it really tough.

Dash is stubborn and determined like me. He wasn't giving up. So I gave him a hint.

This isn't a code where the letters are switched out for other letters. Instead, it's the order you read things that matters.

Oooooooh! I get it!

You don't read from left to right—you read down, then up!

Read the words like this and it will make sense! Just don't get dizzy!

Reading pattern, like a split fence or a wave

If you do that, then the most obvious clue is "BBQ."
Once you spot that, you can find "grill" easily enough.
Then you go back to the start and read:
Go back to Talia's house and look inside the BBQ grill in the backyard!

That's where they'll find the last puzzle pieces, which fit together into the map. Which could itself be tricky since nobody reads maps anymore. Everyone just follows step-by-step directions using GPS.

This is all really great, Talia.

Bet you wish you could be a Mathlete Mermaid!

You know, I kinda do. Even Charles would be impressed by all this.

Wait till he meets Leticia!

I could swear that Dash looked all googly-eyed for just a second. He has a crush on Leticia—I know it! Everyone has a crush on her, starting with Luisa and Skye.

I had one more thing to do. I had to talk to Mr. Wayne. After that, everything would be set for our scavenger hunt pizza party. Everyone is supposed to get here at 11:00 tomorrow, which gives them an hour to solve everything before the pizza gets cold.

A DOORBELL CAN BE A VERY HAPPY SOUND!

DING   DONG   DING   DONG

Just like that first meeting, everyone showed up, all excited and eager to get started. I handed the first clue to Leticia and they were off!

THE THEORY OF RELATIVITY—TIME IS RELATIVE!

When you're waiting for something good to happen, time seems to crawl by. When you're dreading something, time passes all too quickly.

I got so tired of waiting, I lay down in the grass and looked for shapes in the clouds.

I found a goat, chicken, and rabbit.

DEDUCTION #8

THERE ARE DIFFERENT WAYS TO BE GOOD AT THINGS.

I'm good at math because I'm logical and can solve problems. I have a good brain for making deductions, figuring out specific information from general facts. But deductive reasoning also works artistically. At least, I hope it does. Looking at the clouds and finding animals feels like a kind of creative deduction, or maybe it's pattern recognition, which is helpful for figuring out codes. Or maybe it's just a good way to pass the time while waiting.

I saw one of Skye's dragons, the kind she likes to draw.

A Skye dragon in the sky!

Just then, everyone rushed into the backyard. Leo was the first to run in and he was definitely the loudest.

Leo clung to Leticia's hand. He's found a better big sister. Everyone likes her better than me! I didn't want to think about that. This was my party—I made it happen. I shouldn't be thinking about Leticia at all!

The last pieces were scooped out of the BBQ grill, and in a flash, they had the map, marked with a big X.

It had seemed like forever, but it had taken them only 45 minutes—with NO help, no hints. But then, of course, Leticia was in charge.

We all started off for the art room, Leo included. As we left, I could see Dash poking his head over the backyard fence and waving. I gave him a big wave back. This was MY success, too. Not just Leticia's.

Mr. Wayne cheered for us when we came in. He'd decorated the room with streamers.

Go, team, go!

Thank you, Mr. Wayne, and thank you, Talia, for organizing all this.

Yay!

I could feel my cheeks go hot with happiness. Leticia thanked me!

Maybe I'm not a natural leader like Leticia, but I still feel like an important part of our mathlete team. I feel like I belong, something I never felt on Charles's team—or in middle school at all.

And even with Leticia around, I felt like a good big sister, seeing Leo have so much fun, eating pizza with us Mermaids.

OBSERVATION # 39

AND NOW TIME FLIES—
MORE ABOUT HOW TIME IS RELATIVE.

It's only two more days until our first mathlete competition. After all the practice we've done, I should feel ready.
Instead, I feel like I've swallowed a bowling ball, a heavy
weight of math that I can't digest.

Now the competition is tomorrow! I just want it to be over.
I want this horrible feeling in the pit of my stomach to go
away. I want to never compete in anything ever again.

Mom and Dad didn't exactly help the situation.

It sounded like a nightmare! The only thing that could make me more nervous would be them, watching me. Now at least I can fail in peace without their pressure.

Except Leticia will never forgive me. For once in my life, I have a lot of friends. People are asking me about stuff, like codes! I'm about to blow all that.

Tomorrow it will all be over. I'll be back to being the weirdo nobody likes.

"TOMORROW" CAN BE THE BEST WORD OR THE WORST WORD, ALL DEPENDING ON WHAT HAPPENS THE NEXT DAY.

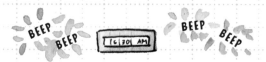

Which will it be for me—best or worst?

My alarm woke me up—tomorrow is here! It's TODAY!

School was miserable. All I could think about was the mathlete competition after school. I couldn't even look at Dash. I could barely talk to Luisa and Skye. I had been so sure I could prove something by winning this competition. Not anymore.

I tried to convince them not to go. It'll be less embarrassing and somehow I'm sure I will fail.

But now, after all my years of pleading, they're all rah-rah for math. Will they still like me once I've shown how bad I am? They've always thought I'm so smart, but what if I'm not, just when it matters the most? I'm counting on them being loyal no matter what! They liked me when I was a weirdo before, so this shouldn't really change anything. At least that's what I keep telling myself.

We lined up for the bus after school. By the time I got there, Leticia was organizing the Mathlete Mermaids behind the boys' team.

She's so good, she even got me excited. I really want to make her proud of me. I want to be the math superhero Skye and Luisa think I can be.

As we got on the bus, all the boys glared at us (except Dash, who gave me a quick wink).

It's weird. We're from the same school, but now we're enemies. Because we all want to win, but only one team can come in first. It had better be us!

The boys started whispering loudly, whispers we were meant to hear.

So Leticia did the same thing, starting lots of eye rolling and loud whispers.

At least the teachers were nice to each other.

OBSERVATION # 41

I ALWAYS THOUGHT I HAD ZERO SCHOOL SPIRIT,
BUT I'M SURPRISED I ACTUALLY DO HAVE A BIT.

I really, really, REALLY wanted to beat Charles, to prove
to him I'm no quitter, that girls can do math. But I have
to admit that if the Mathlete Mermaids don't win, I hope
Charles's team does. At least that way Dash wins—and so
does our school.

I should have been thinking about math, but my mind
wandered. I couldn't help it. I noticed things I shouldn't,
like suddenly everyone's head was a geometrical shape. I
needed to calm my brain down and focus!

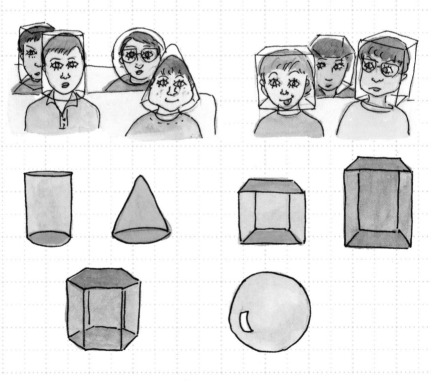

Behave, brain!

I tried taking long, slow, deep breaths. Dad taught me this trick when I had to give my first oral report (terrify-
ing!) and it actually helps. By the time the bus parked, I felt calm and ready to solve any problem. I caught Dash's eye and gave him a big thumbs-up (since winking is another thing I can't do). He nodded the teeniest bit. Which was practically a thumbs-up back.

We got off the bus and filed into the gym where the competition was being held. It smelled like sweaty socks.

DEDUCTION #9

THE FACT IS THERE ARE MORE BOY MATHLETES THAN GIRL ONES.

What can we deduce from that?

 A) Boys are smarter than girls.
 B) Boys are more competitive than girls.
 C) Boys are encouraged to do math and girls aren't.
 D) Boys are encouraged to compete and girls aren't.

I deduce that the answers are C and D, not A and B. Thinking that more boys are mathletes because boys are better at math is an example of confusing cause and effect. The effect of there being more boy mathletes is NOT caused by their being better at math. There are other causes that make that result happen.

WELCOME

But the only way to prove it is for the Mathlete Mermaids to win. Or at least place in the top three teams. That's a tall order.

The gym was set up with tables for one member from each team plus a podium for the quizmaster. There are 10 teams, and one player from each team has a turn to answer a question. Whoever presses the buzzer first gets to answer. If they're wrong, they lose and a kid from an opposing team can press the buzzer. Each right answer gives your team points, and the team with the most points wins.

There are 20 questions, so some girls on our team will be up more than once. Not me, I hope. Taking math tests is one thing, but this was a horrible combination of an oral report and big final exam—a whole lot of pressure with all these people staring at you. It's scary!

Leticia didn't look at all worried. How does she manage to be cool and smart? I thought the two were an impossible combination. There's no way she can be as perfect as she seems.

Just when I should be thinking of math . . .

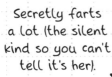

POSSIBLE HIDDEN FLAWS IN LETICIA'S PERFECTNESS

Secretly farts a lot (the silent kind so you can't tell it's her).

Has no idea how to play games with a little brother.

Is good at math, but is a terrible artist.

Has a bad temper, which I just haven't seen yet, but I will.

And maybe, just maybe, she makes mistakes under pressure. Like now. Some of the other girls looked as nervous as I felt. I tried to block out Leticia and the smelly gym stink, and focus only on math.

Looking around and seeing that we were the only girls was a big shock. I shivered as I realized that if I hadn't started the Mathlete Mermaids, I would be the only girl on ANY team. I felt sick.

Hey, are you OK? You look like you've just seen a ghost. Don't worry, you've got this!

Do I? Do I really?

Of course you do!

PEOPLE AREN'T AFRAID OF THE THINGS
YOU THINK WOULD BE THE SCARIEST.

Like more people fear public speaking than shark attacks.
Think about it—more people would rather face a hungry
shark than an audience!

LIST OF FEARS

(from least terrifying to absolutely petrifying)

1. Shark attack
2. Lightning strike
3. Earthquake
4. Zombie apocalypse
5. Public speaking
6. MATHLETE COMPETITIONS

Leticia was cool and calm. I was a wreck. I felt like my legs were made of spaghetti and my feet were giant, clumsy meatballs.

I just couldn't do it. I ran to the bathroom and stared at myself in the mirror, thinking, "This is what a coward looks like."

Then I told myself to stop it and get a grip—just try to answer one problem.

Only one. I can do that. I can do that.

MATHLETE
COMPETITION
TODAY

I'm going to prove that girls can do the math! This team was my idea. I'm a good teammate. I can do it!!

I tiptoed into the hall, heart pounding. Dash came out of the boys' bathroom. He looked the way I felt.

We looked at each other nervously. Neither one of us wanted to go back in right away.

I wasn't sure which was worse—team pressure or my parents.

We walked back to the gym together. Nothing had changed, and everything had changed.

I felt calmer knowing that Dash was as nervous as I was and that he was cheering me on. I finally felt ready. I <u>could</u> do this!

I looked at our score and was excited to see that the Mathlete Mermaids were doing great! Just as I'd hoped! Kayla was answering a problem as I sat down, and she got it right! We were tied with Charles's team for first place.

About time you got here.

You're up next.

Which was actually a good thing—there was no time to get anxious again.

All right, boys—and girl—here we go: Last month I spent $24 on magnets that cost $0.80 each, and this month I spent $24 on magnets that cost $1.20 each. The average cost per magnet is:
A) $0.92 B) $0.96
C) $1.00 or D) $1.04

This was an easy logic problem. It was just a question of who was fastest, so I decided to take a shortcut and guess. The difference between $0.80 and $1.20 is $0.40. Split that in half and you have $0.20. $0.80 + $0.20 = $1.00, and $1.20 - $0.20 = $1.00.

I pressed the buzzer first.

One dollar!

Sorry, that is incorrect.

The announcer didn't look at all sorry.

And that was that. I'd made a stupid mistake. I quickly
did the actual math in my head. The first set of magnets
cost $24 for $0.80 each and 24 divided by .80 = 30, so
that's 30 magnets. The second set of magnets cost $24
and were $1.20 each, with 24 divided by 1.20 equaling 20.
Now I know the total spent is $48 and the total number of
magnets is 50, so 48 divided by 50 = .96. I should have done
the math, not guessed!

But it didn't feel OK. Charles had warned me about shortcuts. But why would I listen to him? And it was all over so fast.

I guess that's what being part of a team means. Feeling even worse when you mess up because you mess up for everyone. Was Charles right after all—I'm just a bad teammate?

Dash was up next. I wanted him to get his answer right more than ever!

Go, Dash!
You've got this!

But he didn't. At least not this time. Sophie answered first. We were tied again!

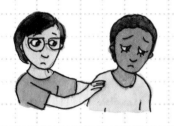

Dash looked miserable as he practically crawled back to his seat. But at least he hadn't answered wrong. He just wasn't fast enough. I wanted to whisper that to him, but, of course, he was sitting far away. I didn't have the chance.

Adrian patted Dash's shoulder. Leroy said something to him, the right kind of something.

I'm glad he has new friends to support him, boys who know he's a good teammate.

A few questions later, we were still tied with Charles's team for first place.

And now, for the last question, let's get all our team captains up here.

This round decides the winner!

May the best team win!

Leticia looked like she was having fun. Charles looked determined. Everyone else looked worried.

The room got very quiet. Mr. Wayne looked proud and surprised and excited all at once. I bet he didn't expect we would do this well. Mr. Douglas also looked proud and surprised and excited. Maybe because no matter which team won, our school would be the winner.

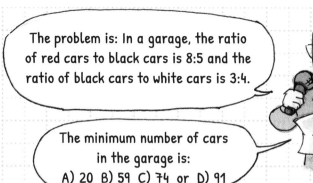

The problem is: In a garage, the ratio of red cars to black cars is 8:5 and the ratio of black cars to white cars is 3:4.

The minimum number of cars in the garage is:
A) 20 B) 59 C) 74 or D) 91

I took out my notebook to do the math, but before I could finish, Charles slammed the buzzer.

Seventy-four!

That is incorrect.

Leticia pressed the buzzer next.

Fifty-nine!

Correct.

The winner of our first mathlete competition of the year is the Mathlete Mermaids!

Then something even more incredible happened. Charles walked up to Leticia and shook her hand.

You're good. Really good.

All the Mathlete Mermaids are.

I had no idea Charles could be so polite. I wondered if I'd been wrong about him. But then . . .

Well, most of them.

Here he glared at me.

We're proud of all our team members.

She is? Proud of me? I didn't believe it but I liked hearing her say it.

Sorry about your problem, Talia. I have to admit I like shortcuts, too. And sometimes they do work.

You just have to know which ones to trust. We should work on that in our next practice.

Work on shortcuts? Really?

Surya's right— there are some good shortcuts. Time for us to share those.

But first we have a celebration to plan!

Dash was already celebrating!

I waited to hear what Charles would say to that. Maybe his own teammates could convince him.

Charles did NOT want to be reminded of that!

OBSERVATION # 43

YOU THINK YOU KNOW WHAT PEOPLE WILL DO, BUT
SOMETIMES THEY'LL SURPRISE YOU—IN A GOOD WAY.

(This is such an important observation, I've made it twice.)

Dash walked right up to me in front of everyone.

Then Adrian and Leroy joined him. And I could see why
they're Dash's friends.

Charles still looked pretty sour, but peer pressure works both ways—it can make you do the right thing as well as the wrong thing. As more and more boys came up to congratulate us, the ones who didn't looked petty. Even boys from other schools congratulated Leticia. One kid even wanted to interview her for his school newspaper!

Charles started out on the edge of the crowd, but I could tell Leticia had convinced him that girls really can do the math.

The bus ride back was totally different from the one going out. The biggest thing was . . .

Dash sat next to me! Right in front of everyone! I didn't feel like a loser anymore.

So we both made mistakes. That felt pretty crummy, but I have to admit, once Charles got his problem wrong, too, mine didn't seem so bad.

I know what you mean! I really enjoyed that, too.

Plus you should be proud you actually answered.

I was too freaked out to even do that.

PUBLIC COMPETITION MAY BE AN EVEN WORSE
FEAR THAN PUBLIC SPEAKING. AND MORE COMMON.

That felt like the real victory. Not the competition, but that Dash and I can be normal friends again, not secret anything.

Once Dash sat next to me, something weird happened. Everyone sat all mixed together, most of the boy math-letes suddenly eager to talk to the girls.

Even Charles sat in the middle of us girls as if he was on our team. Still, there were some holdouts, despite the peer pressure.

By the end of the bus ride, Mr. Douglas and Mr. Wayne looked like best buddies.

How about we join these two teams?

Give the Mermaids a chance to discuss it and you talk it over with your team.

Not everyone might agree.

Oh, I think they will. And I hope the Mermaids will, too.

Mr. Douglas wants a winning team, even if it means including lots of girls. I may have disappointed him, but the Mermaids didn't!

There was a long silence.

YAY!
YES!

Then an explosion of cheers. We're going to celebrate with our rivals? Is that weird?

Dash gave me a big thumbs-up. No, it's not weird at all. It's the way it should have been from the beginning.

Boys and girls can both do math!

Walking home together, straight from the bus, was the best walk ever. I know Dash and I should have been depressed about messing up, but we were too excited by what was happening—boys and girls talking about math together, breaking the strange rules about not being friends.

Dash said he didn't care about Charles and nobody else did, either. The kid on the bus was right—their team needed a new captain. I agreed with that, but it still didn't make sense that the boys would start respecting me. After all, I'd gotten my problem wrong, just like Charles. I reminded Dash of that. But he said I was focusing on the wrong thing.

I was walking on air when I came into the house.

But then I saw my parents.

OBSERVATION # 45

PARENTS SAY THEY CARE ABOUT TEAMWORK, BUT REALLY THEY ONLY CARE ABOUT <u>YOUR</u> WORK.

They started grilling me about my problem, what it was, what I'd done wrong—did I need tutoring, maybe an online course?

But they could have lost because of YOU!

Trust a little brother not to try to spare your feelings.

Mom and Dad didn't say anything more. They didn't have to. Their faces said it all.

Leo followed me into my room.

Leo, it's really OK. Our team won. The boys are talking to us now.

Dash and I can be friends again.

Leo perked up. He knew what really mattered.

You and Dash? Friends again?

Yes, just like before.

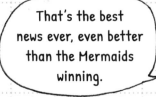

That's the best news ever, even better than the Mermaids winning.

I agree!

I told Leo that the Mermaids would practice shortcuts so I'd know which ones to trust and which ones not to use. I wouldn't make the same mistake again.

You said you had stage fright, too.

Yeah, I got nervous.

But that's something Dash and I are going to work on together, dealing with an audience.

Leo offered right away to be our practice audience. Sure, I agreed. Why not?

I can help!

I know I can!

OBSERVATION # 46

SOMETIMES KIDS HAVE TO TEACH THEIR PARENTS THINGS, NOT THE OTHER WAY AROUND.

Dinner that night was pretty awkward. My parents didn't know what to say. Luckily, Leo did.

It's so great that Dash and Talia are friends again. And you know what? I'm going to help them with the stage fright thing. I'm a big help!

You're a big part of the Mathlete Mermaids.

I'm a member, remember?

Hey, that rhymes!

Member, remember, member, remember!

Leo had a point. But what about Charles? And I wasn't ready to give up the Mermaids so quickly.

Should the Mermaids have only one victory?

On the way to school, all the way there and into the building, Dash and I talked about whether the two teams should join or not. Which was better than talking about how disappointed our parents were.

Dash reminded me that we could be on the same team again. We could help each other. Of course I liked that idea. But . . .

Well, we wouldn't be the Mermaids since only girls are mermaids. We'd need a new name.

But it would still be a special team, one where boys and girls work together.

So the Mathlete Mermaids would disappear entirely? I'm just not sure . . .

At the Mermaid team meeting after school, Leticia said the same things as Dash. She thought the teams would be stronger together than apart. And that working as teammates proved that girls were as good as boys.

You don't worry they'll try to take over everything?

I'd like to see them try.

Yeah, let them try that!

We can handle them!

Who will be our sponsor?

Mr. Wayne or Mr. Douglas?

That's up to them. But I'm staying team captain!

Everyone put their hands up right away, even Skye and Luisa. It was that old peer pressure thing. I hoped my peers were right . . .

So it was unanimous. At least from the Mermaids. We'd have to see what the boys said.

We had just finished the vote when there was a knock on the door.

I'm here to tell you that the mathlete team has voted to join with the Mermaids with Leticia as captain.

We just have one request.

I wasn't surprised but the other Mermaids were.

Leticia answered for all of us.

TEACHERS HAVE FEELINGS, TOO.

You don't think of teachers as people who get sad or upset, but they do. Mr. Wayne was very quiet during our discussion. I wondered what he thought. Maybe he was worried he wouldn't be our sponsor anymore. Maybe he likes being with us. Everyone wants to be part of a winning team.

Mr. Wayne, what do you think of our decision? And do you want to still be our sponsor?

I'm impressed with you all. Some people would have been mad at the boys for not making girls feel welcome.

You set a good example by showing them how to be inclusive.

Was that what we were doing? I thought we just wanted to be friends with boys—well, for me, one boy in particular.

Exactly! They can learn a lot from us!

They definitely can. And I think Mr. Douglas can also learn a lot from you, so I'll step down and leave the teacher sponsor spot to him.

Mr. Wayne said we'd always be welcome in the art room. We can drop by and draw whenever we want.

We'll miss you! You've been great.

You can bet I'll be at all your competitions, cheering you on.

Leticia's party was perfect. (How could it not be since she's so perfect?) She had math bingo and a treasure chest piñata. Plus she had us play charades, using hieroglyphics and emoji codes to spell out what we were supposed to act out.

It was proof positive that boys and girls can have fun together, even if (maybe because?) most of the boys got goo-goo eyed around Leticia.

Hmmm, the boys might have wanted to merge our teams not because of our brains but because of something else entirely. Another reason to thank Leticia, I guess.

But I have to admit it kinda bugged me. Another new thing about middle school . . .

Still, I have to give Leticia credit for all she's done. She's shown me how to be a good teammate. And without her, I wouldn't have Dash as a school friend again. That means a lot!

I can lose personally, but what matters is how the team does. And next time I'll do better. I have to!

I never told you how grateful I am that you were the captain of the Mermaids.

You showed me that I can make mistakes but still work hard for our team. The group matters more than one person.

I really had learned a lot from Leticia.

Thank YOU for starting all this! I didn't know math could be this much fun. We all owe you a lot!

That was just what Dash had said.

And I _can_ take a compliment.

Once all the games had been played and people were eating pizza, Leticia started the discussion about names.

And here I was, thinking Charles was the snake, but that was mean of me. Especially since I'm sure he'd call me the goat.

Chimeras have lion and goat heads (both!), a lion body, eagle wings, and a snake tail.

Really weird!

I'd go for something more simple, like Super Mathletes.

You mean something more boring!

I never thought I'd hear anyone criticize Charles like that. He really has lost some status. I almost felt sorry for him. Almost. We took a vote: Mermaids 7, Dragons 5, Super Mathletes 1, Chimeras 13.

Dash's idea won! Now we're the Chimeras! They're almost as fun to draw as dragons.

Why can't we be people with all kinds of qualities? Like wings, scales, sharp eyes, and strong muscles? The best of everything?

You're absolutely right! I wish I'd thought of that.

You may not have thought it, but you're living that way.

You don't follow the weirdo rules of what girls are "supposed" to be like.

And I don't follow the rules for boys.

And here I thought that was because I was some kind of freak.

It's the rules that are freaky! And we're Chimeras who can ignore them!

I told you Dash has a wonderful way of seeing the world! I can't believe I almost lost him as a friend over stupid freaky rules.
Better to be a Chimera!

RULES MATTER ONLY IF YOU LET THEM.

Dash is 100% right—we get to decide which "rules" of middle school we want to follow. And the rules about how boys and girls are "supposed" to be should definitely to be ignored.

Just imagine if I followed lion rules? Or goat rules?

We'd be enemies to ourselves!

    There are some social codes I want to follow, like being nice to people instead of criticizing them, not talking with my mouth full, not cutting in front of people in line. I have trouble with the rules about what's normal. Because I have to ask, "Normal how?" And who gets to decide what normal is? Girls doing math, for example, is definitely normal, even if the middle-school code doesn't agree!

I STARTED THIS NOTEBOOK WITH THE OBSERVATION THAT MIDDLE SCHOOL IS NOT WHAT I EXPECTED.

That's still true, only in a completely different way from what I meant in Observation #1.

Funny how the same words can mean very different things, even without tone or emojis to add meaning.

Turns out I worried about all the wrong things about middle school and had no idea of what the good things would be. Middle school is not what I expected at all—I never thought I'd be a mathlete, a Mathlete Mermaid, and now a Mathlete Chimera!

I still have to work on the stage fright thing and learn which shortcuts to trust, but I have a whole team to help me. And now I know I can be a good teammate, even if I mess up.

Best of all, I've cracked the code that lets Dash and me still be friends. Even in middle school. All we have to do is be Chimeras!

CODES REALLY ARE EVERYWHERE!

Here's another code Mom told me about—a visual code used by NASA when they sent a rover to Mars. The name of the rover was *Perseverance* (a great name!) and as it landed on the surface of Mars, a red-and-white parachute deployed. Some people noticed that the parachute had an odd pattern, not symmetrical as you would expect. Instead it looked like this:

   Mom wasn't the only computer programmer to see the pattern as binary code. She explained that when she sees things that have only two elements (squares and circles, for example, or in this case, red and white), she thinks of 1s and 0s, the basic elements of computer coding. So each red section became a 1 and each white section a 0.

The result was a long string of 1s and 0s. First Mom tried breaking the 1s and 0s into groups of 8, since that's how a lot of computer code is written. But that still made no sense. A friend was working on the same problem (a lot of computer nerds were!) and she tried groups of 10. That created a pattern where every 10 numerals had three 0s in a row. That was a pattern and the key to solving the code! Each group of 10 was given a number (1, 2, 3, etc.). And each number was then translated into a letter, with 1 = A, 2 = B, and so on.

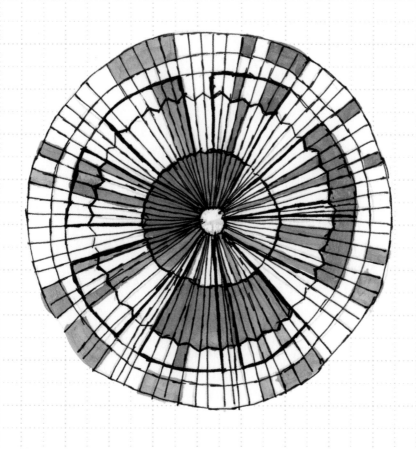

Read the innermost ring first, then the middle, finally the outer, all in a clockwise direction.

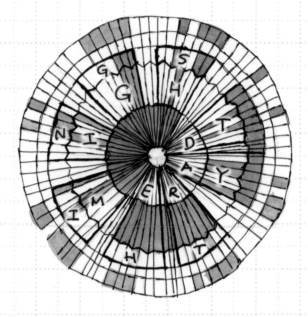

DARE MIGHTY THINGS

I think it's amazing that NASA scientists and engineers took the time to have fun with code on such an important mission. And who knows, maybe an alien will land on Mars and read this message!

I read mighty things!

AUTHOR'S NOTE

This book is based on my own experience as a middle-school mathlete. Like Talia, I was the only girl on the team. But I didn't think to get other girls involved. Instead, I gave up after the first competition when none of my male teammates would speak to me. (It was the longest bus ride ever!) I didn't have the determination to stick it out, and I didn't pursue my interest in math after that. This story is my chance to act—through Talia—the way I wish I had then.

Dash is based on a real person as well, as are the other friends and Talia's parents. When my mom, now in her eighties, was recently hospitalized and I kept her company, she asked me why I'd disappointed my teachers so much by giving up on math. "You were so good at it!" she lamented. I told her I thought I'd done pretty well anyway.

So, thank you to those long-ago math teachers who encouraged me by selecting me as a mathlete and promoting me to higher-level classes at a time when girls were widely considered to be less good at math. Fortunately, that belief has mostly changed, though there's still a lot of catching up to do.

ACKNOWLEDGMENTS

Thank you to my fellow writers who read drafts of this story and helped me hone the characters and language: Gennifer Choldenko, Diane Fraser, Betsy Partridge, Emily Polsby, and Pam Turner. I couldn't have done it without you. I'm also grateful to Susan Van Metre for having faith in this project and to all the folks at Candlewick who helped make it a stronger story. And finally, to all the readers who've written to me over the years saying how my stories resonated with them. We often think we're alone when so many others share our experiences!

ABOUT THE AUTHOR

MARISSA MOSS has written and illustrated more than seventy books, including many in the Amelia's Notebook series. When she wrote the first Amelia book more than twenty-five years ago, the format of a hand-written notebook with art on every page was so novel, editors didn't know what to make of it. They worried that librarians and booksellers wouldn't know how to categorize or shelve it. Now, of course, the notebook format is every-where, especially in the classroom.

Along with creating the Amelia books, Marissa Moss has written historical journals that are currently used in elementary- and middle-school curricula; picture book biographies such as *The Eye That Never Sleeps: How Detective Pinkerton Saved President Lincoln* and *Barbed Wire Baseball*, which won the California Book Award and California Young Reader Medal; and the young adult graphic memoir *Last Things*. Marissa Moss lives in Northern California.

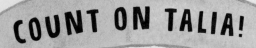

COUNT ON TALIA!

LOOK OUT FOR
THE NEXT
BOOK IN
THE SERIES,
COMING SOON!